M336
Mathematics and Computing: a third-level course

GROUPS & GEOMETRY

UNIT GR2
ABELIAN AND CYCLIC GROUPS

Prepared for the course team by
Bob Coates & Bob Margolis

The Open University

This text forms part of an Open University third-level course.
The main printed materials for this course are as follows.

Block 1
Unit IB1   Tilings
Unit IB2   Groups: properties and examples
Unit IB3   Frieze patterns
Unit IB4   Groups: axioms and their consequences

Block 2
Unit GR1   Properties of the integers
Unit GR2   Abelian and cyclic groups
Unit GE1   Counting with groups
Unit GE2   Periodic and transitive tilings

Block 3
Unit GR3   Decomposition of Abelian groups
Unit GR4   Finite groups 1
Unit GE3   Two-dimensional lattices
Unit GE4   Wallpaper patterns

Block 4
Unit GR5   Sylow's theorems
Unit GR6   Finite groups 2
Unit GE5   Groups and solids in three dimensions
Unit GE6   Three-dimensional lattices and polyhedra

The course was produced by the following team:

Andrew Adamyk (BBC Producer)
David Asche (Author, Software and Video)
Jenny Chalmers (Publishing Editor)
Bob Coates (Author)
Sarah Crompton (Graphic Designer)
David Crowe (Author and Video)
Margaret Crowe (Course Manager)
Alison George (Graphic Artist)
Derek Goldrei (Groups Exercises and Assessment)
Fred Holroyd (Chair, Author, Video and Academic Editor)
Jack Koumi (BBC Producer)
Tim Lister (Geometry Exercises and Assessment)
Roger Lowry (Publishing Editor)
Bob Margolis (Author)
Roy Nelson (Author and Video)
Joe Rooney (Author and Video)
Peter Strain-Clark (Author and Video)
Pip Surgey (BBC Producer)

With valuable assistance from:

Maths Faculty Course Materials Production Unit
Christine Bestavachvili (Video Presenter)
Ian Brodie (Reader)
Andrew Brown (Reader)
Judith Daniels (Video Presenter)
Kathleen Gilmartin (Video Presenter)
Liz Scott (Reader)
Heidi Wilson (Reader)
Robin Wilson (Reader)

The external assessor was:

Norman Biggs (Professor of Mathematics, LSE)

The Open University, Walton Hall, Milton Keynes, MK7 6AA.

First published 1994. Reprinted 1997, 2002 , 2005.

Copyright © 1994 The Open University

All rights reserved. No part of this publication may be reproduced, stored in a retrieval system or transmitted in any form or by any means, without written permission from the publisher or a licence from the Copyright Licensing Agency Limited. Details of such licences (for reprographic reproduction) may be obtained from the Copyright Licensing Agency Ltd of 90 Tottenham Court Road, London, W1P 9HE.

Edited, designed and typeset by the Open University using the Open University TEX System.

Printed in Malta by Gutenberg Press Limited.

ISBN 0 7492 2164 X

This text forms part of an Open University Third Level Course. If you would like a copy of *Studying with the Open University*, please write to the Central Enquiry Service, PO Box 200, The Open University, Walton Hall, Milton Keynes, MK7 6YZ. If you have not already enrolled on the Course and would like to buy this or other Open University material, please write to Open University Educational Enterprises Ltd, 12 Cofferidge Close, Stony Stratford, Milton Keynes, MK11 1BY, United Kingdom.

CONTENTS

Study guide	4
Introduction	5
1 Direct products	6
2 Abelian groups and groups of small orders	13
3 Cyclic groups (audio-tape section)	20
4 Subgroups and quotient groups of cyclic groups	25
5 Direct products of cyclic groups	29
Solutions to the exercises	33
Objectives	44
Index	44

STUDY GUIDE

Section 1 is rather shorter than average. The remaining sections probably require an average amount of study time.

There is an audio programme associated with Section 3 of this unit. That section is fairly theoretical but provides useful practice in using the number-theoretic ideas from *Unit GR1*.

INTRODUCTION

This unit marks the beginning of the main part of the Groups stream.

In this unit we shall begin to build up the body of results that we shall need to prove an important theorem about Abelian groups that is our target. One way of regarding this main theorem is as corresponding to the Unique Prime Factorization Theorem for the integers. In this unit we shall look both at the building blocks (corresponding to the primes) and at the method by which they are combined (corresponding to products).

This target theorem will be proved in Unit GR3.

The seemingly small addition of the Abelian axiom to the group axioms makes an enormous difference to what can be said in general about Abelian groups compared with non-Abelian ones. For our target theorem, we only need results about Abelian groups. However, many of these results either apply in general or have generalizations to non-Abelian groups. Where the proofs generalize fairly easily to the non-Abelian case, we shall either give the general proof or ask you to do the generalization as an exercise.

A word about notation is in order here. When discussing Abelian groups, we shall often use additive notation. Thus we shall write

$$a + b, \quad -a \quad \text{and} \quad na$$

rather than

$$ab, \quad a^{-1} \quad \text{and} \quad a^n.$$

Note that a is an element of the Abelian group concerned but that n is an element of the integers.

For groups which may not be Abelian, we shall stick to multiplicative notation.

In this unit, we continue several themes from *Unit IB4*, particularly 'new groups from old' using subgroups, quotients and direct products and also the use of generators and relations.

Another theme that we shall explore is whether newly constructed groups share the properties of their ancestors. For example, is it true that the direct product of Abelian groups is Abelian; or that the direct product of cyclic groups is cyclic?

Our preoccupation with cyclic groups is partly because they are the simplest examples of Abelian groups and partly because, as we shall see in *Unit GR3*, they turn out to be the building blocks of *all* finite Abelian groups.

Section 1 discusses the conditions under which a group may be written as a direct product of normal subgroups.

In Section 2 we classify all groups of order less than or equal to 8. The results provide a number of useful examples of groups. The methods used indicate how such classifications can be tackled.

As we saw in *Unit IB4*, any free group on one generator is isomorphic to the additive group $\mathbb{Z}$. In Section 3, we shall see that $\mathbb{Z}$ is even more important: we shall show that *every* one-generator Abelian group is isomorphic to a quotient group of $\mathbb{Z}$.

Section 4 discusses quotients and subgroups of cyclic groups.

Finally, in Section 5, we prove our main result, a decomposition theorem for expressing finite cyclic groups as direct products of smaller cyclic groups.

1 DIRECT PRODUCTS

In the Groups stream we want, eventually, to be able to 'break down' any group by expressing it in terms of 'simpler' components. We have already seen, in *Unit IB4*, how groups can be 'built up' from other groups by using the direct product construction. We now want to see, in this section, under what conditions a group is isomorphic to a direct product of two of its subgroups. The usefulness of this is that knowledge of the factors in such a product gives information about the product. For example, as we saw in *Unit GR1*, the order of an element in a direct product of groups is related to the orders of its components.

If $g \in G$ has order m and $h \in H$ has order n, then $(g, h) \in G \times H$ has order $\operatorname{lcm}\{m, n\}$.

To illustrate this idea, consider the group of symmetries of the rectangle,

$$\Gamma(\square) = \{e, r, h, v = rh\},$$

where r is the half-turn about the centre and h and v are the reflections in the horizontal and vertical axes of symmetry. The Cayley table of $\Gamma(\square)$ is as follows:

	e	r	h	v
e	e	r	h	v
r	r	e	v	h
h	h	v	e	r
v	v	h	r	e

We shall show how $\Gamma(\square)$ can be represented as a direct product of two subgroups simpler than $\Gamma(\square)$, i.e. we shall show that $\Gamma(\square)$ is *isomorphic* to the direct product of two such subgroups.

We have already seen, in *Unit IB4*, that $\Gamma(\square)$ is isomorphic to the Klein group, V.

By Lagrange's Theorem, the only possible orders for subgroups of $\Gamma(\square)$ are 1, 2 and 4. We aim to express $\Gamma(\square)$ in the form

$$\Gamma(\square) = A \times B,$$

where A and B are subgroups of $\Gamma(\square)$. Now, by the definition of direct products of *any* two groups A and B, $|A \times B| = |A||B|$. Therefore if either A or B has order 1, then the other must have order 4 and is the whole group. This would not be expressing $\Gamma(\square)$ as a product of simpler groups. Thus, it only makes sense to try and express $\Gamma(\square)$ as the direct product of two subgroups of order 2. There are three subgroups of order 2. We pick the following two:

$$H_1 = \{e, r\} \quad \text{and} \quad H_2 = \{e, h\}.$$

The other is

$$H_3 = \{e, v\}.$$

In fact, any two of the three subgroups would do just as well.

The Cayley table for $H_1 \times H_2$ is:

	(e,e)	(r,e)	(e,h)	(r,h)
(e,e)	(e,e)	(r,e)	(e,h)	(r,h)
(r,e)	(r,e)	(e,e)	(r,h)	(e,h)
(e,h)	(e,h)	(r,h)	(e,e)	(r,e)
(r,h)	(r,h)	(e,h)	(r,e)	(e,e)

Since we have $rh = v$ in $\Gamma(\square)$, the Cayley table for $\Gamma(\square)$ can be written in the following way:

	e	r	h	rh
e	e	r	h	rh
r	r	e	rh	h
h	h	rh	e	r
rh	rh	h	r	e

As these two tables have identical forms, it follows that the function ϕ defined by

$$\phi : H_1 \times H_2 \to \Gamma(\square)$$
$$(x, y) \mapsto xy$$

is an isomorphism. Thus $\Gamma(\square)$ can be represented as, i.e. is isomorphic to, the direct product of two of its subgroups. Since each of these subgroups is cyclic of order 2, we can also say

$$\Gamma(\square) \cong \mathbb{Z}_2 \times \mathbb{Z}_2 \cong C_2 \times C_2.$$

Not all attempts at such decompositions of groups as direct products of subgroups will work, as the following example shows.

Example 1.1

Consider the permutation group S_3, which has six elements:

$$S_3 = \{e, (12), (13), (23), (123), (132)\}.$$

If S_3 could be written as a *non-trivial* direct product, i.e. as a direct product in which neither subgroup is of order 1, it would have to be as the direct product of subgroups of orders 2 and 3.

The group S_3 does possess subgroups of orders 2 and 3. For example,

$$H_1 = \{e, (12)\}$$

is of order 2 and

$$H_2 = \{e, (123), (132)\}$$

is of order 3.

However, any subgroup of order 2 is cyclic and is isomorphic to $\mathbb{Z}_2$. Equally, any subgroup of order 3 is cyclic and is isomorphic to $\mathbb{Z}_3$.
It follows, therefore, that *any* attempt to form a direct product of such subgroups leads to a group isomorphic to

$$\mathbb{Z}_2 \times \mathbb{Z}_3,$$

which you proved (in *Unit IB4*) is not isomorphic to S_3. ◆ $\mathbb{Z}_2 \times \mathbb{Z}_3 \cong \mathbb{Z}_6$ is Abelian; S_3 is not.

The example of S_3 shows that, even if subgroups of suitable orders exist, the group may not be isomorphic to a direct product of these subgroups.

In fact, there are precise conditions under which a group is isomorphic to a direct product of subgroups. We can determine these conditions by extending some of the work in *Unit IB4*. We do so in two stages: first we establish necessary conditions on the subgroups and then show that these conditions are also sufficient.

These conditions form the content of Theorem 1.1.

We start by showing that if a group G is the direct product of two of its subgroups, then there are restrictions on the subgroups.

This discussion will be the proof of the first half of Theorem 1.1.

Assume, therefore, that H_1 and H_2 are subgroups of a group G and assume that ϕ defined by

$$\phi : H_1 \times H_2 \to G$$
$$(h_1, h_2) \mapsto h_1 h_2$$

is an isomorphism. We consider, in turn, the consequences of ϕ being onto, one–one and satisfying the morphism property.

Onto

The image set of ϕ consists of all products

$$h_1 h_2$$

of an element of H_1 and an element of H_2. Hence, as ϕ is onto, every element of G is expressible in this form. Using the notation

$$H_1 H_2 = \{h_1 h_2 : h_1 \in H_1, h_2 \in H_2\},$$

this says that we must have

$$G = H_1 H_2.$$

In other words, every element of G is the product of an element of H_1 with an element of H_2. Summing up, we have the following.

> If ϕ is an isomorphism, then
> $$G = H_1 H_2.$$

One–one

Assume that
$$\phi((h_1, h_2)) = \phi((k_1, k_2)).$$
Then the one–one condition on ϕ implies that
$$(h_1, h_2) = (k_1, k_2),$$
which, by the definition of ordered pair, implies that
$$h_1 = k_1 \quad \text{and} \quad h_2 = k_2.$$
We now interpret this result, using the definition of ϕ.
Since $\phi((h_1, h_2)) = \phi((k_1, k_2))$ means that
$$h_1 h_2 = k_1 k_2,$$
the fact that ϕ is one–one means that
$$h_1 h_2 = k_1 k_2 \quad \Rightarrow \quad h_1 = k_1 \text{ and } h_2 = k_2.$$
Thus, because ϕ is one–one, each element of G can be expressed as a *unique* product of the form $h_1 h_2$ with $h_1 \in H_1$ and $h_2 \in H_2$.

Exercise 1.1

Show that the one–one condition implies that
$$H_1 \cap H_2 = \{e\}.$$

Hint Use the fact that $h = he = eh$ for any $h \in G$.

The result of the previous exercise tells us the following.

> If ϕ is an isomorphism, then
> $$H_1 \cap H_2 = \{e\}.$$

Morphism property

For any two elements (h_1, h_2) and (k_1, k_2) of the direct product, the morphism property of ϕ gives

$$\phi((h_1, h_2)(k_1, k_2)) = \phi((h_1, h_2))\phi((k_1, k_2))$$
$$\Rightarrow \quad \phi((h_1 k_1, h_2 k_2)) = (h_1 h_2)(k_1 k_2)$$
$$\Rightarrow \quad h_1 k_1 h_2 k_2 = h_1 h_2 k_1 k_2$$
$$\Rightarrow \quad k_1 h_2 = h_2 k_1 \quad \text{(using the left and right cancellation rules).}$$

What this means is that every element of H_1 commutes with every element of H_2.

We now show that we can deduce that H_1 is a *normal* subgroup of G.
To do so, we have to show that, for each $a \in G$,
$$a H_1 a^{-1} \subseteq H_1.$$

Suppose that $a \in G$. Then, by our first result above, we can write
$$a = h_1 h_2,$$
where $h_1 \in H_1$ and $h_2 \in H_2$.

Now, any element of aH_1a^{-1} is of the form aha^{-1}, where $h \in H_1$. So
$$\begin{aligned} aha^{-1} &= (h_1h_2)h(h_1h_2)^{-1} \\ &= h_1h_2hh_2^{-1}h_1^{-1} \\ &= h_1hh_2h_2^{-1}h_1^{-1} \quad \text{(since } h \in H_1 \text{ and } h_2 \in H_2 \text{ commute)} \\ &= h_1hh_1^{-1} \in H_1 \quad \text{(since } h, h_1, h_1^{-1} \in H_1\text{).} \end{aligned}$$

Thus
$$aH_1a^{-1} \subseteq H_1$$
as required.

We now ask you to show that H_2 is also a normal subgroup of G.

Exercise 1.2

Prove that, for any $a \in G$,
$$aH_2a^{-1} \subseteq H_2.$$

Combining the two normality proofs, we have the following.

> If ϕ is an isomorphism, then both H_1 and H_2 are normal subgroups of G.

Summing up, if ϕ, as defined above, is an isomorphism, then all three of the following conditions are satisfied:

(a) $G = H_1H_2$;

(b) $H_1 \cap H_2 = \{e\}$;

(c) H_1 and H_2 are normal subgroups of G.

This completes the proof of the first half of the following theorem.

> **Theorem 1.1 Internal direct product theorem**
>
> If H_1 and H_2 are subgroups of a group G, then
> $$\begin{aligned} \phi : H_1 \times H_2 &\to G \\ (h_1, h_2) &\mapsto h_1h_2 \end{aligned}$$
> is an isomorphism if and only if all three of the following conditions hold:
>
> (a) $G = H_1H_2$;
>
> (b) $H_1 \cap H_2 = \{e\}$;
>
> (c) H_1 and H_2 are normal subgroups of G.
>
> When these three conditions are satisfied, then
> $$G \cong H_1 \times H_2$$
> and we say that G is the **internal direct product** of H_1 and H_2.

The direct product is 'internal' in that H_1 and H_2 are subgroups of G.

Proof

As remarked above, we have proved that the conditions are necessary, i.e. we have proved that, if ϕ is an isomorphism, then the three conditions hold. For sufficiency, we need to prove that, if the three conditions hold, then ϕ is an isomorphism.

The three conditions, in turn, give the onto, one–one and morphism properties of ϕ.

Onto

Since $G = H_1 H_2$, every $g \in G$ can be written as

$$g = h_1 h_2, \quad h_1 \in H_1, \; h_2 \in H_2.$$

Hence

$$g = \phi((h_1, h_2))$$

and ϕ is onto.

One–one

Suppose that

$$\phi((h_1, h_2)) = \phi((k_1, k_2)).$$

Then

$$h_1 h_2 = k_1 k_2, \quad h_1, k_1 \in H_1, \; h_2, k_2 \in H_2.$$

So

$$k_1^{-1} h_1 = k_2 h_2^{-1}.$$

But the left-hand side is an element of H_1 and the right-hand side is an element of H_2. Since both sides are equal, they belong to both H_1 and H_2 and, hence, to the intersection $H_1 \cap H_2$.

However,

$$H_1 \cap H_2 = \{e\}.$$

Thus

$$k_1^{-1} h_1 = k_2 h_2^{-1} = e.$$

It immediately follows that

$$h_1 = k_1 \quad \text{and} \quad h_2 = k_2,$$

and so

$$(h_1, h_2) = (k_1, k_2).$$

Thus ϕ is one–one.

Morphism property

This is rather more involved. We first prove that the normality of H_1 and H_2, combined with the fact that they have trivial intersection, implies that every element of H_1 commutes with every element of H_2.

Let $h_1 \in H_1$ and $h_2 \in H_2$. Now, a and b commute if and only if $aba^{-1}b^{-1}$ is the identity. Since we wish to prove that h_1 and h_2 commute, we consider the element x of G given by

$$\begin{aligned} x &= h_1 h_2 h_1^{-1} h_2^{-1} \\ &= (h_1 h_2 h_1^{-1}) h_2^{-1}. \end{aligned}$$

Because $h_2 \in H_2$ and H_2 is normal, we have

$$h_1 h_2 h_1^{-1} \in H_2.$$

If a and b commute, then $ab = ba$. Multiplying by inverses, on the right, gives

$$aba^{-1}b^{-1} = e.$$

These steps are reversible, hence justifying the statement.

Thus,
$$x = (h_1 h_2 h_1^{-1}) h_2^{-1} \in H_2.$$

Similarly, we can write
$$x = h_1 (h_2 h_1^{-1} h_2^{-1}).$$

Because $h_1^{-1} \in H_1$ and H_1 is normal, we have
$$h_2 h_1^{-1} h_2^{-1} \in H_1.$$

Thus,
$$x = h_1 (h_2 h_1^{-1} h_2^{-1}) \in H_1.$$

We have shown that $x \in H_1 \cap H_2$, and so $x = e$. From
$$x = h_1 h_2 h_1^{-1} h_2^{-1} = e$$
we can immediately deduce that
$$h_1 h_2 = h_2 h_1.$$

Now we can tackle the final part of the proof.

Suppose that (h_1, h_2) and (k_1, k_2) are elements of $H_1 \times H_2$. Then
$$\begin{aligned}
\phi((h_1, h_2)(k_1, k_2)) &= \phi((h_1 k_1, h_2 k_2)) \\
&= h_1 k_1 h_2 k_2 \\
&= h_1 h_2 k_1 k_2 \quad \text{(by the commutativity just proved)} \\
&= \phi((h_1, h_2)) \phi((k_1, k_2)).
\end{aligned}$$
∎

This theorem gives another reason why S_3 is not the direct product of two subgroups isomorphic to $\mathbb{Z}_2$ and $\mathbb{Z}_3$. For, although
$$H_2 = \{e, (123), (132)\}$$
is a normal subgroup of order 3, S_3 has no normal subgroup of order 2.

Exercise 1.3

Consider the cyclic group
$$C_6 = \{e, a, a^2, a^3, a^4, a^5\}.$$

Show that C_6 has normal subgroups of orders 2 and 3 which satisfy the conditions of Theorem 1.1. Deduce that
$$C_6 \cong C_2 \times C_3.$$

Before we move on, one more theorem about direct products of groups is going to be useful.

Theorem 1.2

If A, B and C are groups, then:
(a) $A \times B \cong B \times A$;
(b) $A \times (B \times C) \cong (A \times B) \times C$.

Proof

(a) Consider the function ϕ defined by:
$$\phi : A \times B \to B \times A$$
$$(a, b) \mapsto (b, a)$$

We want to show that ϕ is an isomorphism.

One–one

If
$$\phi(a_1, b_1) = \phi(a_2, b_2),$$
then
$$(b_1, a_1) = (b_2, a_2).$$

Hence, by the definition of ordered pairs,
$$b_1 = b_2 \quad \text{and} \quad a_1 = a_2,$$
and so
$$(a_1, b_1) = (a_2, b_2)$$
and ϕ is one–one.

Onto

If (b, a) is any element of the codomain $B \times A$, then $b \in B$ and $a \in A$, so $(a, b) \in A \times B$ and
$$\phi(a, b) = (b, a).$$

Hence ϕ is onto.

Morphism property

Let (a_1, b_1) and (a_2, b_2) be any two elements of $A \times B$. Then
$$\phi((a_1, b_1)(a_2, b_2)) = \phi(a_1 a_2, b_1 b_2)$$
$$= (b_1 b_2, a_1 a_2)$$
$$= (b_1, a_1)(b_2, a_2)$$
$$= \phi(a_1, b_1)\phi(a_2, b_2).$$

This completes the proof that $A \times B \cong B \times A$.

(b) We ask you to prove the second part of the theorem in the following exercise. ∎

Exercise 1.4

Show that the function ψ defined by
$$\psi : A \times (B \times C) \to (A \times B) \times C$$
$$(a, (b, c)) \mapsto ((a, b), c)$$

is an isomorphism.

The solution to Exercise 1.4 completes the proof that
$A \times (B \times C) \cong (A \times B) \times C$. ■

The first part of Theorem 1.2 says that if we alter the order of groups in a direct product then we obtain an isomorphic group.

The second part of Theorem 1.2 says that the bracketing of terms in a direct product of three (or more) groups is unnecessary.

So, just as for, say, the multiplication of integers, these 'commutative' and 'associative' laws mean that a direct product of two or more groups may be written in any order, without the need for brackets, since all such expressions produce isomorphic groups.

2 ABELIAN GROUPS AND GROUPS OF SMALL ORDERS

In this section we consider some of the immediate consequences of adding the Abelian axiom to the standard group axioms.

Definition 2.1 Abelian group

A group G is Abelian if

$$xy = yx$$

for all $x, y \in G$.

For Abelian groups, the group operation will often be written $+$ so the above condition becomes

$$x + y = y + x.$$

When $+$ is used for the group operation, the identity will be written as 0, the inverse of x as $-x$ and the result of adding n copies of x as nx.

We shall still use the notation G/H for quotient groups of additive Abelian groups, but the left coset of H containing x will be written

$$x + H.$$

Note that nx is not a group product, but the additive equivalent of the nth power of x.

A number of concepts about groups simplify considerably for Abelian groups. The following exercises consider some consequences of the definition.

Exercise 2.1

Prove that every cyclic group is Abelian.

Exercise 2.2

Prove that all subgroups of Abelian groups are Abelian.

Exercise 2.3

Prove that all subgroups of an Abelian group are normal.

A consequence of this result is that, in Abelian groups, corresponding left and right cosets are equal.

Exercise 2.4

Prove that, if H is a subgroup of the Abelian group G, then the quotient group G/H is Abelian.

Exercise 2.5

Prove that the direct product of two groups G_1 and G_2 is Abelian if and only if both G_1 and G_2 are Abelian.

Only one of the results that we have just asked you to prove was 'if and only if'. For the others, we are going to show that the converses do not hold by finding simple counterexamples.

For convenience, we list the 'one-way' results that you have just proved:

(a) Every cyclic group is Abelian.
(b) Every subgroup of an Abelian group is Abelian.
(c) Every subgroup of an Abelian group is normal.
(d) Every quotient group of an Abelian group is Abelian.

When you suspect that a result may only hold one way round, you should look for a counterexample to the converse. When looking for counterexamples, it is best to start with simple ones; in the case of results about groups, it is best to start with groups of small order.

We look for counterexamples to the converses of these results, in turn.

(a) It is not true that every Abelian group is cyclic. The simplest counterexample is V, the Klein group. This is Abelian, has four elements, but no element of order 4, as the three non-identity elements are all of order 2.

Remember that $V \cong \Gamma(\square)$.

(b) It is true that, if *all* subgroups of a group are Abelian, the whole group must be, since it is a subgroup of itself.

A more interesting question is whether a group all of whose *proper* subgroups are Abelian must itself be Abelian. This time the answer is no, and the simplest counterexample is provided by the smallest non-Abelian group, namely the group D_3 of symmetries of an equilateral triangle. All of its proper subgroups are cyclic, with 1, 2 or 3 elements, and are therefore Abelian, whereas D_3 is non-Abelian.

A *proper subgroup* is one which is not the whole group.

Recall, from *Unit IB2*, that D_3 and S_3 are isomorphic.

(c) It is not true that a group in which every subgroup is normal must be Abelian. We shall discuss a group providing a counterexample a little later in this section.

(d) It is true that, if *every* quotient group of a group G is Abelian, then so is G. This is because one normal subgroup is the trivial subgroup $\{e\}$ and $G/\{e\}$ is isomorphic to G.

A more interesting question is whether, if every quotient of a group by a *non-trivial* normal subgroup is Abelian, then the group must be Abelian. The answer is no, and we ask you to verify, in the next exercise, that S_3 is a counterexample.

A *non-trivial subgroup* is one different from $\{e\}$.

Exercise 2.6

(a) Prove that the non-Abelian group S_3 has only two non-trivial normal subgroups.

(b) By considering the orders of the corresponding quotient groups, show that the quotient groups are Abelian.

In the remainder of this section we shall classify all groups of order up to 8. By this we mean that we shall give a 'standard' set of non-isomorphic groups one for each of the orders $1, \ldots, 8$. We shall show that every group whose order does not exceed 8 is isomorphic to a group in our standard set.

The process of classifying such groups will give you practice in using the basic properties of groups and subgroups. It will also provide a useful stock of potential counterexamples for future use.

For orders $1, \ldots, 7$ we have done much of the work already. The work for order 8 gives some indication that classifying groups of a particular order is, in general, quite difficult. However, for Abelian groups of any prescribed order, a complete classification is known and will be discussed in *Unit GR3*.

We begin with an exercise that provides a useful result.

Exercise 2.7

Let G be a group in which every non-identity element has order 2. Show that G is Abelian.

Hint For any $x, y \in G$, consider x^2, y^2 and $(xy)^2$.

Some orders have already been disposed of. We record the results here for completeness.

Order 1

There is only one group of order 1, namely $\{e\}$. By this we mean that our standard group of order 1 is $\{e\}$.

Order 2, 3, 5 and 7

These are prime orders. They are covered by the following theorem from *Unit GR1*.

> **Theorem**
>
> Let p be a prime number. Then there is only one group of order p, namely the cyclic group of order p.

This is Theorem 4.1 from *Unit GR1*.

Hence there is only one group of each of the orders 2, 3, 5 and 7: the cyclic group of that order. Our standard groups of these orders are C_2, C_3, C_5 and C_7.

We might equally well have chosen $\mathbb{Z}_2$, $\mathbb{Z}_3$, $\mathbb{Z}_5$ and $\mathbb{Z}_7$.

Order 4

If the group has an element of order 4, then it is cyclic, and we take C_4 as the standard group.

If there is no element of order 4, then, by Lagrange's Theorem, all three non-identity elements have order 2. By Exercise 2.7, the group is Abelian. The group must be of the form

$$\{e, a, b, c\},$$

where

$$a^2 = b^2 = c^2 = e.$$

By closure, $ab = ba$ is in the group.

Using the left and right cancellation rules, we can show that ab cannnot be e, a or b (as you may wish to check for yourself).
Hence,

$$ab = ba = c.$$

Similarly,

$$bc = cb = a \quad \text{and} \quad ac = ca = b.$$

These are precisely the relations that hold in the Klein group, V.

Hence there are only two possible groups of order 4:

$$C_4 \quad \text{and} \quad V \cong C_2 \times C_2.$$

We saw that $V \cong C_2 \times C_2$ in Section 1.

We take C_4 and $C_2 \times C_2$ as the standard groups.

Order 6

We already know that there are at least two non-isomorphic groups of order 6, the cyclic group C_6 and the dihedral group D_3. In fact these are the only two.

Remember that $C_6 \cong \mathbb{Z}_6$ and $D_3 \cong S_3$, and that $\mathbb{Z}_6 \not\cong S_3$.

We shall look in detail at the proof of this result since it uses some of our previous results and since the methods will be used again to deal with the groups of order 8.

The overall strategy is a proof by exhaustion, considering the possible orders of the elements of the group.

Suppose that G is a group of order 6. Firstly, by Lagrange's Theorem, the only possible orders for the elements of G are 1, 2, 3, and 6. If there is an element of order 6 then the group G is cyclic.

We now show that, if there is *no* element of order 6 in G, then G is D_3.

We show that G has an element of order 3. As a consequence, it will follow that G also has an element of order 2.

We eliminate the possibility that *all* non-identity elements of G have order 2. If this were the case then, by Exercise 2.7, G would be Abelian. Then, if a and b are any two non-identity elements of G, as in the case above for order 4, they generate a subgroup

$$\{e, a, b, ab\}$$

of order 4, which contradicts Lagrange's Theorem.

It is easy to check that this is a subgroup.

We know, therefore, that G has an element a of order 3, which generates a subgroup

$$H = \{e, a, a^2\}$$

of order 3.

Now let b be an element of G not in H. Because b is not in H, Theorem 3.2 of *Unit GR1* tells us that the right cosets Hb and $He = H$ are disjoint. Hence, the remaining elements of G are given by the right coset

$$Hb = \{b, ab, a^2b\}.$$

We now consider the possibilities for the element $ba \in G$.

It happens to be convenient to use the *right* coset here. The argument could be carried out using the left coset but it would be less easy to recognize the final results.

Exercise 2.8

By eliminating the other possibilities, prove that ba must be either ab or a^2b.

Exercise 2.9

Show that b has order 2.

Exercise 2.10

Show that, if $ba = ab$, then ab has order 6.

So, $ab = ba$ contradicts the fact that G does not contain an element of order 6. Hence, if G is not cyclic, then we know that $ba = a^2b$.

Thus G is generated by a and b, with relations

$$a^3 = b^2 = e, \quad ba = a^2b.$$

This is equivalent to the description of D_3 given in the Appendix to *Unit IB4*, if you replace a by r and b by s.

This description would have been less easily recognizable if we had used the left coset bH earlier.

We have, therefore, that G is either

$$C_6 \cong C_2 \times C_3 \quad \text{or} \quad D_3 \cong S_3.$$

We take C_6 and S_3 as our standard groups of order 6.

We saw that $C_6 \cong C_2 \times C_3$ in Exercise 1.3.

Order 8

Our approach is similar to that for order 6.

By Lagrange's Theorem, the orders of the elements of G divide 8. We look at the two extreme cases first, since they turn out to be the easiest.

If there is an element of order 8 in G, then G is the cyclic group C_8.

If all non-identity elements are of order 2, then G is Abelian. Consider any two non-identity elements, a and b, say. Then, as in the order 4 and order 6 cases, these two elements generate a subgroup

$$H = \{e, a, b, ab\}$$

of order 4. Furthermore, all of its elements are of order 2. Therefore, as we can deduce from our work in Section 1, it is isomorphic to the Klein group. Thus

$$H \cong V \cong C_2 \times C_2.$$

Since $ab = ba$, $(ab)^2 = a^2b^2 = ee = e$.

Let c be an element of G not in H. Then,

$$K = \langle c \rangle \cong C_2$$

is a subgroup of G, such that

$$H \cap K = \{e\}.$$

Furthermore, since the right cosets $He = H$ and Hc are disjoint, and contain four elements each, we have

$$G = HK.$$

Also, since all subgroups of an Abelian group are normal, H and K are normal.

Hence, by Theorems 1.1 and 1.2,

$$G \cong H \times K \cong C_2 \times C_2 \times C_2.$$

We have dealt with the case where there is an element of order 8 and the case where there are no elements of order 8 or 4. It remains to deal with the case where there is an element of order 4, but no element of order 8.

Let us now assume that G contains an element a of order 4, but no element of order 8. The element a generates a subgroup

$$A = \{e, a, a^2, a^3\}$$

of order 4. Since A has index 2 in G it is normal in G. *By Exercise 3.4 of Unit IB4.*

Let b be an element of G which is not in A. The subgroup A has two left cosets in G, namely A itself and

$$bA = \{b, ba, ba^2, ba^3\}.$$

Hence a and b generate G.

We shall now show, by considering relations, that there are only three possible groups generated by a and b.

Since $b \neq e$ and there is no element of order 8 in G, b has order 2 or 4. In either case

$$b^4 = e.$$

Exercise 2.11 _____

(a) By considering the product $(bA)(bA)$ in the quotient group

$$G/A \cong C_2,$$

show that

$$b^2 \in A.$$

(b) Use the result from part (a) to show that *Hint* $(b^2)^2 = b^4 = e.$

$$b^2 = e \quad \text{or} \quad b^2 = a^2.$$

As in the case of groups of order 6, we now consider the element ab.

This element belongs to the right coset Ab and, by the normality of A, to the left coset bA. Hence ab must be one of

$$b, \ ba, \ ba^2 \text{ or } ba^3.$$

Exercise 2.12 _____

Show that ab cannot be b or ba^2. *Hint* Consider $a^2 b = a(ab).$

We thus have four combinations of possibilities for the values of b^2 and of ab, namely:

$b^2 = e, \quad ab = ba;$
$b^2 = e, \quad ab = ba^3;$
$b^2 = a^2, \quad ab = ba;$
$b^2 = a^2, \quad ab = ba^3.$

We shall show that two of these four possibilities lead to the same group. Each of the other two leads to a different group, thus giving three further groups of order 8, in addition to C_8 and $C_2 \times C_2 \times C_2$, making a grand total of five.

We begin with the case $ab = ba$, which means that G is Abelian (since a and b generate G). We show that there is an element c of order 2 not in A. We show that both options for b^2, namely e or a^2, lead to the same group.

If $b^2 = e$, we take $c = b \neq e$. If $b^2 = a^2 \neq e$, we take $c = ba$ and then we have

If a group G is generated by two elements a and b such that $ab = ba$, then G is Abelian. (You may wish to check this result for yourself.)

$$\begin{aligned} c^2 &= (ba)^2 \\ &= b^2 a^2 \quad \text{(since } ab = ba\text{)} \\ &= a^2 a^2 \\ &= a^4 \\ &= e. \end{aligned}$$

So in both cases $c^2 = e$.

Consider the subgroup C, where

$$C = \langle c \rangle,$$

which is normal because G is Abelian.

We know that A is normal, and we have

$$A \cap C = \{e\}.$$

Furthermore, as in the case of H and K above,

$$G = AC.$$

Hence, by Theorems 1.1 and 1.2,

$$\begin{aligned} G &\cong A \times C \cong C_4 \times C_2 \\ &\cong C \times A \cong C_2 \times C_4. \end{aligned}$$

We take $C_2 \times C_4$ as our standard form.

We have now found three Abelian groups of order 8. No two are isomorphic because C_8 has an element of order 8, the other two do not; $C_2 \times C_4$ has an element of order 4, which $C_2 \times C_2 \times C_2$ does not.

The next possibility

$$G = \langle a, b : a^4 = e,\ b^2 = e,\ ab = ba^3 \rangle$$

is our standard description of D_4, the symmetries of the square, where a is anticlockwise rotation through $\pi/2$ about the centre (usually denoted by r) and b is a reflection in an axis of symmetry (usually denoted by s). Since this group is non-Abelian (as $ab \neq ba$) it cannot be isomorphic to any of the three groups of order 8 that we have found so far.

We can certainly write down the remaining possibility:

$$G = \langle a, b : a^4 = e,\ b^2 = a^2,\ ab = ba^3 \rangle.$$

From Section 5 of *Unit IB4*, we know that there *exists* a group with these generators and relations. What we do *not* yet know, however, is whether this group has 8 elements, and if it does, whether it is or is not isomorphic to one of our other four groups of order 8.

In the following exercise, we ask you to show that two given 2×2 matrices, with entries from the complex numbers $\mathbb{C}$, generate a group of order 8 and satisfy the above relations. We also ask you to show that this group is not isomorphic to any of the other four groups of order 8.

Exercise 2.13

Consider the matrices
$$a = \begin{bmatrix} i & 0 \\ 0 & -i \end{bmatrix}, \quad b = \begin{bmatrix} 0 & -1 \\ 1 & 0 \end{bmatrix}, \quad \text{where } i^2 = -1.$$

(a) Show that a and b satisfy the relations
$$a^4 = e, \quad b^2 = a^2 \quad \text{and} \quad ab = ba^3.$$

(b) Show that the eight elements
$$e, a, a^2, a^3, b, ba, ba^2, ba^3$$
are all distinct.

(c) Show that the set
$$\{e, a, a^2, a^3, b, ba, ba^2, ba^3\}$$
is closed under matrix multiplication by writing out its Cayley table. Deduce that the set, under matrix multiplication, is a group of order 8.

(d) Show that this group is not isomorphic to any of the other four groups of order 8.

In this representation of the group, the identity e is the 2×2 identity matrix $\mathbf{I}$.

Hint To show that it is not isomorphic to D_4, consider the number of elements of order 2.

The matrix group in Exercise 2.13 is a concrete example of the group with general description
$$G = \langle a, b : a^4 = e, \ b^2 = a^2, \ ab = ba^3 \rangle$$
given above. This group, in this abstract form in terms of generators and relations, is often denoted by Q and is called Hamilton's quaternion group. It is the promised example of a non-Abelian group all of whose subgroups are normal (as you may check if you wish).

Strictly speaking, this group should be denoted by Q_8 rather than Q. This is because it is the first of a family of two-generator groups defined by similar relations.

Summing up, the three Abelian groups together with D_4 and Q constitute a complete classification of groups of order 8.

We now tabulate our findings about groups of order up to 8.

Order	Group(s)
1	C_1
2	C_2
3	C_3
4	$C_4, C_2 \times C_2$
5	C_5
6	$C_6 \cong C_2 \times C_3, S_3 \cong D_3$
7	C_7
8	$C_8, C_2 \times C_4, C_2 \times C_2 \times C_2, D_4, Q$

It is worth noting that $C_1 \cong S_1$, $C_2 \cong S_2 \cong D_1$ and $C_2 \times C_2 \cong D_2$, as you can check by reference to the definitions in the Appendix to Unit IB4.

If we list the *Abelian* groups from the catalogue of groups up to order 8, we obtain the following.

Order	Abelian group(s)
1	C_1
2	C_2
3	C_3
4	$C_4, C_2 \times C_2$
5	C_5
6	$C_6 \cong C_2 \times C_3$
7	C_7
8	$C_8, C_2 \times C_4, C_2 \times C_2 \times C_2$

This last table gives some indication of why we are interested in cyclic groups in particular. In fact the main theorem in the Groups stream of the course is an easily stated generalization of the results about Abelian groups in the table.

3 CYCLIC GROUPS (AUDIO-TAPE SECTION)

In this section we shall give a complete description of all cyclic groups. We do so by showing that any cyclic group is isomorphic to one of a list of concrete examples of groups. In fact the concrete examples will be

$$(\mathbb{Z}, +) \quad \text{and} \quad (\mathbb{Z}_n, +_n),$$

for each positive integer n.

To make our starting point clear, we remind you that we have defined a group G to be cyclic if G is generated by a single element, that is

$$G = \langle a \rangle = \{a^k : k \in \mathbb{Z}\}.$$

Note that, in this section, we retain multiplicative notation for G, despite the fact that we know that cyclic groups are Abelian.

The cyclic group with which you are most familiar is $\mathbb{Z}$, which is generated by the element 1.

$\mathbb{Z}$ is also generated by -1.

You should now listen to the audio programme for this unit, referring to the tape frames below when asked to during the programme.

1 Aim

Prove theorem

If G is a cyclic group

Then G is isomorphic

         either to $(\mathbb{Z}, +)$

         or to $(\mathbb{Z}_n, +_n)$ for suitable n

> side effect: this will give a *definition* of $\mathbb{Z}_n$

2 Cyclic groups

Definition   If G is cyclic then
$$G = \langle a \rangle = \{a^k : k \in \mathbb{Z}\} \quad \text{(multiplicative)}$$
$$= \{ka : k \in \mathbb{Z}\} \quad \text{(additive)}$$

Examples

Rotation group: $\langle r[\alpha] \rangle = \{(r([\alpha])^k : k \in \mathbb{Z}, \alpha = 2\pi/5\}$ (finite)

Even integers: $\langle 2 \rangle = \{2k : k \in \mathbb{Z}\}$ (infinite)

3 Strategy

Start with $G = \langle a \rangle = \{a^k : k \in \mathbb{Z}\}$

Exploit association $k \mapsto a^k$ (homomorphism?)

Use *First Isomorphism Theorem*   (Theorem 4.5 of *Unit IB4*)

Get G isomorphic to quotient of $\mathbb{Z}$

Then find all quotients of $\mathbb{Z}$

4 Homomorphism and onto

Define $\phi : \mathbb{Z} \to G$
$$k \mapsto a^k$$

Want   (a) ϕ to be a *homomorphism*

            that is, $\phi(k + l) = \phi(k)\phi(l)$

      (b) ϕ to be *onto*

            that is, $\operatorname{Im}(\phi) = G$

Exercise 3.1   Prove these statements

4A

Solution 3.1

(a) $\phi(k+l) = a^{k+l}$ — (definition of ø)
$= a^k a^l$ — (rule of indices)
$= \phi(k)\phi(l)$ — (definition of ø)

So ø is a *homomorphism*

(b) G is cyclic, so every element of G is a power of a

That is, if $x \in G$, $x = a^k$ for some k

Then $\phi(k) = a^k = x$ — (definition of ø)

So ø is *onto*

5

Consequences

$\phi : \mathbb{Z} \to G$ is a homomorphism

So $\operatorname{Im}(\phi) \cong \mathbb{Z}/\ker(\phi)$ — (First Isomorphism Theorem)

(ø is *onto* so $\operatorname{Im}(\phi) = G$)

So $G \cong \mathbb{Z}/\ker(\phi)$ — (What can ker(ø) be?)

6

Kernels

Kernels are normal subgroups, and vice versa

(Theorems 4.4 and 4.6 of *Unit IB4*)

Question   What are the normal subgroups of $\mathbb{Z}$?

Simplify   All subgroups of $\mathbb{Z}$ are normal — ($\mathbb{Z}$ is Abelian)

Revised problem   What are the subgroups of $\mathbb{Z}$?
What are the corresponding quotient groups?

($G \cong \mathbb{Z}/$(a subgroup of $\mathbb{Z}$))

7

Trivial case

$\mathbb{Z}$ has the trivial subgroup $\{0\}$

What is the quotient group $\mathbb{Z}/\{0\}$?

Exercise 3.2

(a) Show that each left coset contains just one element
(b) Show that $\mathbb{Z}$ and $\mathbb{Z}/\{0\}$ are isomorphic

7A

Solution 3.2

(a) A typical left coset is $k + \{0\}$,
which is $\{k\}$, so has only one element

(b) Need to establish
$$\psi : \mathbb{Z} \mapsto \mathbb{Z}/\{0\}$$
$$k \to k + \{0\}$$

is one–one *(natural homomorphism)*
(natural homomorphisms are always onto)

If $\quad \psi(k) = \psi(l)$
then $\quad k + \{0\} = l + \{0\}$
i.e. $\quad \{k\} = \{l\}$
and $\quad k = l$

(usual proof of one-one-ness)

Therefore ψ is an *isomorphism*

So $\mathbb{Z} \cong \mathbb{Z}/\{0\}$

8

Non-trivial cases

$\mathbb{Z}$ cyclic, so subgroup H must also be cyclic

So H generated by n, say
$$H = \langle n \rangle = \{kn : k \in \mathbb{Z}\}$$

Choices

If $n = 0$, then $H = \{0\}$ *(already done in Frame 7)*
So deal with cases when $n \neq 0$

Further simplify

n and $-n$ generate same subgroup
So deal only with $n > 0$

9

New notation

If H is a non-trivial subgroup of $\mathbb{Z}$ then
$$H = \{kn : k \in \mathbb{Z}, n > 0\}$$

Notation: $n\mathbb{Z}$ for set of integer multiples of n

If H is non-trivial subgroup, then it is $n\mathbb{Z}$ for some $n > 0$

10 Progress so far

(a) Used $\phi : \mathbb{Z} \to G$
    $k \mapsto a^k$ *(ø is homomorphism, onto)*

   to show $\text{Im}(\phi) = G \cong \mathbb{Z}/\ker(\phi)$ *(First Isomorphism Theorem)*

(b) Used facts about $\mathbb{Z}$ *(cyclic, Abelian)*

   to show $\ker(\phi)$ same as subgroup of $\mathbb{Z}$

Remaining task Describe $\mathbb{Z}/n\mathbb{Z}$ for $n > 0$ *(every subgroup of $\mathbb{Z}$ can be written as $n\mathbb{Z}$)*

11 Quotient groups $\mathbb{Z}/n\mathbb{Z}$

Know Quotient is *cyclic* *($\mathbb{Z}$ is cyclic)*

   Generator is $n\mathbb{Z} + 1$ *(1 generates $\mathbb{Z}$)*

Question How many elements?

   $1 + n\mathbb{Z}, 2 + n\mathbb{Z}, \ldots, n + n\mathbb{Z}$ *($= n\mathbb{Z}$)*

Distinct Difference of two from
   $0, \ldots, n-1$ *($n \in n\mathbb{Z}$)*
   can't be multiple of n

Conclusion $\mathbb{Z}/n\mathbb{Z}$ is cyclic with n elements

12 Summing up

Every cyclic group is isomorphic to a quotient group of $\mathbb{Z}$:

   one *infinite* quotient group, isomorphic to $\mathbb{Z}$
   one *finite* quotient group $\mathbb{Z}/n\mathbb{Z}$ for each $n > 0, n \in \mathbb{Z}$

> So every cyclic group is isomorphic to $\mathbb{Z}$ or to $\mathbb{Z}/n\mathbb{Z}$ for some integer $n > 0$

Observation

Adding $0 + n\mathbb{Z}, 1 + n\mathbb{Z}, \ldots, (n-1) + n\mathbb{Z}$ is like addition modulo n

Can *formally* define $\mathbb{Z}_n = \mathbb{Z}/n\mathbb{Z}$ *(we write the operation on $\mathbb{Z}_n$ as $+_n$)*

> So every cyclic group is isomorphic to $\mathbb{Z}$ or to $\mathbb{Z}_n$ for some integer $n > 0$

We summarize the major result from the tape as follows.

> **Theorem 3.1 Classification of cyclic groups**
>
> (a) Every cyclic group is isomorphic to a quotient group of $(\mathbb{Z}, +)$.
>
> (b) All infinite cyclic groups are isomorphic to $\mathbb{Z}$.
>
> (c) A finite cyclic group of order n is isomorphic to the quotient group $\mathbb{Z}/n\mathbb{Z}$, which is, by definition, $(\mathbb{Z}_n, +_n)$.

On route to proving Theorem 3.1, we proved the following classification of all the subgroups of $\mathbb{Z}$.

> **Theorem 3.2 Subgroups of $\mathbb{Z}$**
>
> The subgroups of $\mathbb{Z}$ are precisely the sets of the form
>
> $$n\mathbb{Z} = \{nk : k \in \mathbb{Z}\},$$
>
> for some non-negative integer n.

Note that, if n is a positive integer, then $n\mathbb{Z}$ is isomorphic to $\mathbb{Z}$ itself.

4 SUBGROUPS AND QUOTIENT GROUPS OF CYCLIC GROUPS

We have already seen, in Section 2 of this unit, that subgroups and quotient groups of Abelian groups are Abelian. A natural question arises as to whether the same is true if 'Abelian' is replaced by 'cyclic'. The answer is yes, as we shall see in this section.

Although we know, from Section 3, that cyclic groups can be written in a specific form, namely as $\mathbb{Z}$ or $\mathbb{Z}_n$, it is sometimes convenient to view them in a more abstract way and in multiplicative notation. We shall adopt this more abstract view in this section.

We shall begin by looking at quotients of cyclic groups.

When we discussed quotients of $\mathbb{Z}$, we found that the quotient group $\mathbb{Z}/n\mathbb{Z}$ is generated by the coset

$$1 + n\mathbb{Z},$$

the coset containing the generator of the original group, $\mathbb{Z}$. The following theorem confirms that this observation generalizes.

> **Theorem 4.1 Quotients of cyclic groups**
>
> Let G be a cyclic group generated by a and let H be a subgroup of G. Then the quotient group G/H is cyclic and generated by the coset aH.

As cyclic groups are Abelian, all of their subgroups are normal (see results (a) and (c) on page 13) and so we can always form the quotient group G/H.

Exercise 4.1

Prove Theorem 4.1.

Now we turn our attention to subgroups.

For the *infinite* cyclic group, that is $\mathbb{Z}$, we already know, from Theorem 3.2, precisely what the subgroups are. They are all of the form

$$n\mathbb{Z}, \quad n \in \mathbb{Z}, n \geq 0.$$

Each of these is a cyclic group generated by n.

We shall now show that every subgroup of a finite cyclic group is also cyclic. In fact, we shall do rather more. As we have seen, Lagrange's Theorem states that, for finite groups, the order of a subgroup must divide the order of the group. For finite *cyclic* groups we can prove a strong converse to Lagrange's Theorem. For every divisor of the order of the group, not only is there a subgroup having that number of elements, but also this subgroup is unique.

The theorem that we shall prove is stated formally as follows.

Theorem 4.2 Subgroups of cyclic groups

Let

$$G = \langle a : a^n = e \rangle$$

be a finite cyclic group of order n.

(a) Every subgroup of G is cyclic.

(b) If q is a factor of n with $n = mq$, then G has a unique subgroup of order q which is generated by the element a^m.

Proof

We tackle the proof in several stages, some of which we ask you to carry out as exercises. An important tool in the proof will be the Quotient–remainder Theorem from *Unit GR1*.

We first deal with the trivial subgroup $\{e\}$. It is cyclic, since it is generated by e. This proves part (a) of the theorem for the trivial subgroup.

We can also deal immediately with the trivial factor, 1, of n. There is a subgroup of order 1, the trivial subgroup, and this is the unique subgroup of this order. Also, $n = n \times 1$ and the trivial subgroup is generated by the element $a^n = e$. This completes the proof of part (b) of the theorem for the trivial factor.

Now we prove part (a) for a non-trivial subgroup.

Let H be a non-trivial subgroup of G. We show that H is cyclic.
Each element of H must be of the form a^k with $0 \leq k < n$.
Among the elements of H, choose the one with the smallest exponent and let this exponent be m. Since H is non-trivial, m is positive.
We now ask you to show that

$$H = \langle a^m \rangle. \qquad \square$$

Exercise 4.2

(a) Suppose that a^k is an element of H. Show that m divides k and hence that

$$H = \langle a^m \rangle.$$

Hint Use the Quotient–remainder Theorem.

(b) Deduce that m divides n.

(c) If $n = mq$, show that the order of H is q.

Proof of Theorem 4.2 continued

We can summarize the results that we have obtained so far as follows. If

$$G = \langle a : a^n = e \rangle$$

is a finite cyclic group of order n and H is a subgroup of G then:

- H is cyclic;
- H is generated by a^m, where m is the smallest positive exponent of a such that $a^m \in H$;
- m divides n;
- if $n = mq$ then H has order q.

Thus we have completed the proof of part (a) of Theorem 4.2.

We now tackle part (b), looking first of all at the *existence* of subgroups whose orders correspond to factors of the order of the group.

By Lagrange's Theorem, the order q of H is a factor of n, the order of G. In fact, we have shown more, namely that H is generated by the element a^m where $n = mq$. This suggests that *given* a factor q of n, the subgroup generated by a^m, where $n = mq$, will have order q.

We now ask you to show that this is true. □

Exercise 4.3

Let

$$G = \langle a : a^n = e \rangle$$

be a finite cyclic group of order n and let q be a factor of n with

$$n = mq.$$

Show that the subgroup

$$H = \langle a^m \rangle$$

has order q.

Proof of Theorem 4.2 continued

Finally, we show that the subgroup corresponding to the factor q of n is *unique*.

Suppose that $n = mq$.
Then we know that there *is* a subgroup

$$\langle a^m \rangle$$

of order q.
Let H be *any* subgroup of order q.
From the proof of part (a), we know that H is cyclic and generated by a^l, where l is the least positive exponent of a occurring among the elements of H.
Furthermore, the order of H is k, where

$$n = lk.$$

But the order of H is q, so $k = q$ and

$$mq = n = lk = lq,$$

and, therefore, $m = l$.
Thus H is the subgroup generated by a^m.

This completes the proof of Theorem 4.2. ■

Theorem 4.2 can be regarded as a strong converse to Lagrange's Theorem for cyclic groups. We use 'strong' because, not only is there a subgroup for *every* factor, but, in addition, the corresponding subgroup is *unique*.

Later in the Groups stream we shall investigate the effect of relaxing the assumptions about the group G. We shall first consider the case where G is finite and Abelian (instead of cyclic) and then go on to relax the condition further to merely finite. At each stage we shall still obtain some sort of converse to Lagrange's Theorem; but, as the conditions on G are relaxed, the conclusions become weaker, although still useful.

We conclude this section by interpreting the results for our standard finite cyclic groups: the groups $\mathbb{Z}_n$.

At this point, a word is in order about how we think of $\mathbb{Z}_n$. We now have the reassurance of a formal definition (from Section 3): $\mathbb{Z}_n$ *is* the quotient group $\mathbb{Z}/n\mathbb{Z}$. However, with this background established, in practice we always think of $\mathbb{Z}_n$ as the set

$$\{0, 1, \ldots, n-1\},$$

with addition carried out modulo n.

Since the order of $\mathbb{Z}_n$ is n, by Theorem 4.2 there is a subgroup corresponding to each factor of n. That is, if q is a factor of n, then $\mathbb{Z}_n$ has a cyclic subgroup of order q. This subgroup must (by Theorem 3.2) be isomorphic to $\mathbb{Z}_q$.

We also know, by Theorem 4.2, a generator of the subgroup. Since $\mathbb{Z}_n$ is generated by 1, the subgroup of order q is generated by

$$m = \frac{n}{q}.$$

Exercise 4.4

(a) Write down the orders of the subgroups of $\mathbb{Z}_{24}$.

(b) For each such subgroup:

   (i) state to which $\mathbb{Z}_k$ it is isomorphic;

   (ii) by using the remark above, about how we think of the groups $\mathbb{Z}_n$, give an explicit list of its elements.

The interpretation of our results about quotient groups is also quite straightforward.

For each factor q of n, there is a subgroup of $\mathbb{Z}_n$ of order q. This subgroup is normal. The corresponding quotient group is cyclic (by Theorem 4.1) and has order

$$\frac{|\mathbb{Z}_n|}{q} = \frac{n}{q}.$$

Hence, the quotient is isomorphic to $\mathbb{Z}_{n/q}$.

5 DIRECT PRODUCTS OF CYCLIC GROUPS

In this section we shall apply the direct product ideas from Section 1 of this unit, and also of Section 1 of *Unit IB4*, to cyclic groups.

Firstly, we note that, irrespective of whether the groups are cyclic or not, the order of the direct product of finite groups is the product of the orders of the individual groups.

This is actually a statement about the number of elements in the Cartesian product of two finite sets.

Secondly, if the groups concerned are cyclic, then they are Abelian and, hence, by Exercise 2.5, so is their direct product.

On the other hand, we also know that direct products of cyclic groups are not *always* cyclic. For example, in Section 1, we showed that

$$\Gamma(\square) \cong \mathbb{Z}_2 \times \mathbb{Z}_2,$$

and $\Gamma(\square)$ is not cyclic.

However, the direct product of cyclic groups may *sometimes* be cyclic. For example,

$$\mathbb{Z}_2 \times \mathbb{Z}_3$$

is an Abelian group with six elements and so must be $\mathbb{Z}_6$, which is cyclic. Alternatively, we could argue that the order of $(1,1)$ in $\mathbb{Z}_2 \times \mathbb{Z}_3$ is the LCM of the order of 1 in $\mathbb{Z}_2$ and the order of 1 in $\mathbb{Z}_3$. Since these orders are 2 and 3, with LCM 6, $\mathbb{Z}_2 \times \mathbb{Z}_3$ has an element $(1,1)$ of order 6 and so is cyclic.

We showed in Section 2 that the only Abelian group of order 6 is $C_6 \cong C_2 \times C_3$.

So, a direct product of cyclic groups may or may not be cyclic. The aim of this section is to determine under what circumstances the direct product

$$\mathbb{Z}_m \times \mathbb{Z}_n$$

is cyclic.

Exercise 5.1

Show that both of the following direct products are cyclic:

(a) $\mathbb{Z}_3 \times \mathbb{Z}_5$;

(b) $\mathbb{Z}_4 \times \mathbb{Z}_5$.

We now ask you to consider examples where the direct product is not cyclic.

Exercise 5.2

(a) Show that $(1,1)$ does not generate $\mathbb{Z}_2 \times \mathbb{Z}_4$.
   Why is this insufficient to prove that $\mathbb{Z}_2 \times \mathbb{Z}_4$ is not cyclic?

(b) Show that $\mathbb{Z}_2 \times \mathbb{Z}_4$ is not cyclic.

In the solution to Exercise 5.2, we saw that the maximum order for an element of the direct product is 4. We could show this directly as follows. If (a,b) is any element of the direct product $\mathbb{Z}_2 \times \mathbb{Z}_4$, then

$$4(a,b) = (4a, 4b)$$
$$= (0,0).$$

Exercise 5.3

Show that the maximum order of any element of $\mathbb{Z}_6 \times \mathbb{Z}_8$ is at most 24 and, hence, that the direct product is not cyclic.

We now ask you to make decisions about some direct products.

Exercise 5.4

For each of the following direct products, decide whether or not it is cyclic and justify your conclusion:

(a) $\mathbb{Z}_4 \times \mathbb{Z}_6$;

(b) $\mathbb{Z}_2 \times \mathbb{Z}_9$.

Inspecting the examples above suggests that the direct product of cyclic groups *of coprime orders* is cyclic, but if the orders are not coprime then the direct product is not cyclic. This is the content of the following theorem.

> **Theorem 5.1 Direct products of cyclic groups**
>
> The direct product
>
> $$\mathbb{Z}_m \times \mathbb{Z}_n$$
>
> of the cyclic groups $\mathbb{Z}_m$ and $\mathbb{Z}_n$ is cyclic if and only if m and n are coprime positive integers.

Proof

If

We prove that, if m and n are coprime, then the direct product is cyclic.

We verify that $(1,1)$ is a generator by showing that it has order mn, which is the order of the direct product.

The element 1 in $\mathbb{Z}_m$ has order m and the element 1 in $\mathbb{Z}_n$ has order n. As m and n are coprime, we have

$$\text{hcf}\{m,n\} = 1.$$

However, by Theorem 2.2 of *Unit GR1*, since m and n are positive,

$$\text{hcf}\{m,n\} \times \text{lcm}\{m,n\} = mn.$$

Hence $\text{lcm}\{mn\} = mn$.

The order of the element $(1,1)$ in the direct product is, therefore, mn. However, the direct product has mn elements and so is cyclic.

We may deduce that

$$\mathbb{Z}_m \times \mathbb{Z}_n \cong \mathbb{Z}_{mn}.$$

Only if

The second half of the proof, which shows that if m and n are not coprime then the direct product is not cyclic, is set as an exercise. ∎

Exercise 5.5

Show that, if m and n are not coprime, no element in the direct product $\mathbb{Z}_m \times \mathbb{Z}_n$ can have order greater than

$$\text{lcm}\{m,n\},$$

and hence that $\mathbb{Z}_m \times \mathbb{Z}_n$ is not cyclic.

Proof of Theorem 5.1 continued

The solution to Exercise 5.5 completes the proof of the theorem. ∎

The proof of the theorem yields the result that, if m and n are coprime, then $\mathbb{Z}_m \times \mathbb{Z}_n \cong \mathbb{Z}_{mn}$. Reading this equivalence from right to left gives the following lemma.

> **Lemma 5.1**
>
> Let m and n be coprime positive integers. Then
> $$\mathbb{Z}_{mn} \cong \mathbb{Z}_m \times \mathbb{Z}_n.$$

This lemma is an example of the sort of decomposition result that is our main concern in the Groups stream. It leads to the main decomposition theorem for cyclic groups, which we prove later in this section.

Exercise 5.6

Show that
$$\mathbb{Z}_{90} \cong \mathbb{Z}_2 \times \mathbb{Z}_9 \times \mathbb{Z}_5.$$

In Exercise 5.6, had we decomposed 90 in a different manner, taking
$$90 = 9 \times 10,$$
for example, we would have obtained
$$\mathbb{Z}_{90} \cong \mathbb{Z}_9 \times \mathbb{Z}_2 \times \mathbb{Z}_5.$$

Other decompositions of 90 produce corresponding direct products. The fact that all the resulting direct product decompositions are isomorphic follows from Theorem 1.2.

Exercise 5.7

Show that, if n is a positive integer with prime decomposition
$$n = p_1^{k_1} \ldots p_r^{k_r},$$
where $p_1 < \cdots < p_r$ are (distinct) primes and $k_1, \ldots, k_r$ are positive integers, then
$$\mathbb{Z}_n \cong \mathbb{Z}_{n_1} \times \cdots \times \mathbb{Z}_{n_r},$$
where
$$n_i = p_i^{k_i}, \quad i = 1, \ldots, r.$$

Exercise 5.8

Prove that
$$\mathbb{Z}_{154} \times \mathbb{Z}_{20} \times \mathbb{Z}_5 \quad \text{and} \quad \mathbb{Z}_{55} \times \mathbb{Z}_{28} \times \mathbb{Z}_{10}$$
are isomorphic.

The solution to Exercise 5.7 provides a decomposition theorem for finite cyclic groups, which we restate as follows.

> **Theorem 5.2 Decomposition of finite cyclic groups**
>
> If n is a positive integer with prime decomposition
> $$n = p_1^{k_1} \ldots p_r^{k_r},$$
> where $p_1 < \ldots < p_r$ are (distinct) primes and $k_1, \ldots, k_r$ are positive integers, then
> $$\mathbb{Z}_n \cong \mathbb{Z}_{n_1} \times \cdots \times \mathbb{Z}_{n_r},$$
> where
> $$n_i = p_i^{k_i}, \quad i = 1, \ldots, r.$$

If we combine the above theorem with the uniqueness of prime decomposition in $\mathbb{Z}$, we see that the decomposition into cyclic groups of prime power order is unique.

See Theorem 5.1 of *Unit GR1*.

Since each prime power order is a factor of the order of the group, then by Theorem 4.2 each $\mathbb{Z}_{n_i}$ is (isomorphic to) a subgroup. Thus the decomposition theorem expresses the group as a direct product of (normal) subgroups.

All subgroups of cyclic groups are normal.

We now have a complete description of all cyclic groups.

- Firstly, we have a list of familiar, concrete, cyclic groups, $\mathbb{Z}$ and $\mathbb{Z}_n$ for each positive n. Every cyclic group is isomorphic to one group in this list.

- Secondly, every finite cyclic group can be written as a direct product in which every term is a cyclic group of prime power order. These cyclic groups of prime power order cannot be written as direct products of cyclic groups of smaller orders. Thus the cyclic groups of prime power order form a collection of fundamental building blocks from which all finite cyclic groups can be constructed.

The fact that cyclic groups of prime power order cannot be written as direct products of cyclic groups of smaller orders is a consequence of Theorem 5.1.

- Thirdly, we know all about the subgroups of a cyclic group.
  For the infinite cyclic group, $\mathbb{Z}$, the subgroups are precisely the sets

  $$n\mathbb{Z} = \{nx : x \in \mathbb{Z}\},$$

  for every non-negative integer n.
  A finite cyclic group has a unique cyclic subgroup corresponding to each factor of its order.

As, later in the Groups stream, we relax our restrictions on the group, from finite cyclic to finite Abelian and then to just finite, we shall aim to prove similar results. That is, we shall look for information about subgroups and decomposition into direct products of simpler groups.

SOLUTIONS TO THE EXERCISES

Solution 1.1

Assume that $h \in H_1 \cap H_2$. We can express h in two ways as an element of $H_1 H_2$, namely

$$h = he, \quad h \in H_1,\ e \in H_2,$$

and

$$h = eh, \quad e \in H_1,\ h \in H_2.$$

Hence

$$\phi((h,e)) = \phi((e,h))$$

and the one–one condition gives

$$(h, e) = (e, h).$$

So, by the definition of ordered pair,

$$h = e.$$

Thus the only element in $H_1 \cap H_2$ is the identity, as required.

Solution 1.2

As before, we write

$$a = h_1 h_2,$$

where $h_1 \in H_1$ and $h_2 \in H_2$.

Let aha^{-1} be any element of aH_2a^{-1}, where $h \in H_2$. Then

$$\begin{aligned}
aha^{-1} &= (h_1 h_2) h (h_1 h_2)^{-1} \\
&= h_1 h_2 h h_2^{-1} h_1^{-1} \\
&= h_1 (h_2 h h_2^{-1}) h_1^{-1} \\
&= h_1 h' h_1^{-1} \quad \text{(where } h' = h_2 h h_2^{-1} \in H_2\text{)} \\
&= h' h_1 h_1^{-1} \quad \text{(since } h_1 \in H_1 \text{ and } h' \in H_2 \text{ commute)} \\
&= h' \in H_2,
\end{aligned}$$

and so

$$aH_2 a^{-1} \subseteq H_2.$$

Solution 1.3

We can define the subgroups as follows:

$$\begin{aligned}
H_1 &= \langle a^3 \rangle = \{e, a^3\}; \\
H_2 &= \langle a^2 \rangle = \{e, a^2, a^4\}.
\end{aligned}$$

Note that $H_1 \cong C_2$ and $H_2 \cong C_3$.

It is clear that e, a^2, a^3, a^4, a^5 are all in $H_1 H_2$. If we note that $a^3 a^4 = a^7 = a$, we see that

$$C_6 = H_1 H_2.$$

Next, by inspection,

$$H_1 \cap H_2 = \{e\}.$$

Lastly, we must deal with the question of normality. Actually, it is easy to prove the stronger result that every element of H_1 commutes with every element of H_2, since everything is a power of a.

Hence, by Theorem 1.1,

$$C_6 \cong C_2 \times C_3.$$

Solution 1.4

One–one

Suppose that
$$\psi(a_1,(b_1,c_1)) = \psi(a_2,(b_2,c_2)).$$
By the definition of ψ,
$$((a_1,b_1),c_1) = ((a_2,b_2),c_2).$$
By the definition of ordered pairs,
$$(a_1,b_1) = (a_2,b_2) \quad \text{and} \quad c_1 = c_2.$$
Applying the definition of ordered pairs again,
$$a_1 = a_2 \quad \text{and} \quad b_1 = b_2.$$
Hence
$$(a_1,(b_1,c_1)) = (a_2,(b_2,c_2)).$$
This completes the proof that ψ is one–one.

Onto

If $((a,b),c)$ is any element of the codomain, then $a \in A, b \in B, c \in C$, and so $(a,(b,c)) \in A \times (B \times C)$ and
$$\psi(a,(b,c)) = ((a,b),c).$$
This completes the proof that ψ is onto.

Morphism property

Let $(a_1,(b_1,c_1))$ and $(a_2,(b_2,c_2))$ be any two elements of the domain. Then
$$\begin{aligned}
\psi((a_1,(b_1,c_1))(a_2,(b_2,c_2))) &= \psi(a_1a_2,((b_1,c_1)(b_2,c_2))) \\
&= \psi(a_1a_2,(b_1b_2,c_1c_2)) \\
&= ((a_1a_2,b_1b_2),c_1c_2) \\
&= ((a_1,b_1)(a_2,b_2),c_1c_2) \\
&= ((a_1,b_1),c_1)((a_2,b_2),c_2) \\
&= \psi((a_1,(b_1,c_1)))\psi((a_2,(b_2,c_2))).
\end{aligned}$$
This completes the proof.

Solution 2.1

Let G be a cyclic group generated by the element g. If x and y are any two elements of G, then, for some integers m and n, we have
$$x = g^m \quad \text{and} \quad y = g^n.$$
Then,
$$\begin{aligned}
xy &= g^m g^n \\
&= g^{m+n} \quad &&\text{(by the rule of indices)} \\
&= g^{n+m} \quad &&\text{(since addition in } \mathbb{Z} \text{ is commutative)} \\
&= g^n g^m \quad &&\text{(by the rule of indices)} \\
&= yx.
\end{aligned}$$
Hence G is Abelian.

If you anticipated the result and used additive notation for G, your proof should read something like the following.

Let G be a cyclic group generated by the element g. If x and y are any two elements of G, then, for some integers m and n, we have

$$x = mg \quad \text{and} \quad y = ng.$$

Then,

$$\begin{aligned} x + y &= mg + ng \\ &= (m+n)g \\ &= (n+m)g \quad \text{(since addition in } \mathbb{Z} \text{ is commutative)} \\ &= ng + mg \\ &= y + x. \end{aligned}$$

Hence G is Abelian.

Solution 2.2

Using additive notation, if H is a subgroup of the Abelian group G then

$$x + y = y + x$$

for all $x, y \in G$ and, as H is a subset of G, certainly

$$x + y = y + x$$

for all $x, y \in H$.

Solution 2.3

Suppose that H is a subgroup of the Abelian group G and that $a \in G$.

The definition of normality $(aH = Ha)$ when written additively becomes $a + H = H + a$.

Any element of $a + H$ can be written

$$a + h, \quad \text{for some } h \in H.$$

But G is Abelian, so

$$a + h = h + a.$$

This tells us that every element of $a + H$ is an element of $H + a$. So

$$a + H \subseteq H + a.$$

Similarly

$$H + a \subseteq a + H,$$

and equality follows. Hence H is a normal subgroup of G.

Solution 2.4

Firstly, we observe that the G/H always exists because, by the previous exercise, any subgroup is normal.

Now, let

$$X = x + H, \quad Y = y + H,$$

where $x, y \in G$, be any two elements of the quotient group G/H. Then, using additive notation in anticipation,

$$\begin{aligned} X + Y &= (x + H) + (y + H) \\ &= (x + y) + H \quad \text{(by the definition of a quotient group)} \\ &= (y + x) + H \quad \text{(since } G \text{ is Abelian)} \\ &= (y + H) + (x + H) \quad \text{(by the definition of a quotient group)} \\ &= Y + X. \end{aligned}$$

Thus G/H is Abelian.

Solution 2.5

Firstly, assume that both G_1 and G_2 are Abelian. Now let

$$x = (a_1, a_2) \quad \text{and} \quad y = (b_1, b_2)$$

be elements of the direct product $G_1 \times G_2$, where

$$a_1, b_1 \in G_1 \quad \text{and} \quad a_2, b_2 \in G_2.$$

Then, using additive notation for the direct product in anticipation of the result,

$$\begin{aligned} x + y &= (a_1 + b_1, a_2 + b_2) &&\text{(by the definition of a direct product)} \\ &= (b_1 + a_1, b_2 + a_2) &&\text{(since } G_1 \text{ and } G_2 \text{ are Abelian)} \\ &= y + x &&\text{(by the definition of a direct product)}. \end{aligned}$$

Secondly, assume that $G_1 \times G_2$ is Abelian. Suppose that a_1 and b_1 are elements of G_1 and that a_2 and b_2 are elements of G_2. Then (a_1, a_2) and (b_1, b_2) are elements of the direct product. Since the direct product is Abelian, we have

$$(a_1, a_2) + (b_1, b_2) = (b_1, b_2) + (a_1, a_2).$$

Hence,

$$(a_1 + b_1, a_2 + b_2) = (b_1 + a_1, b_2 + a_2).$$

From the definition of ordered pairs, this gives

$$a_1 + b_1 = b_1 + a_1 \quad \text{and} \quad a_2 + b_2 = b_2 + a_2.$$

Thus, G_1 and G_2 are Abelian.

Solution 2.6

(a) The elements of S_3 are

$$\{e, (12), (13), (23), (123), (132)\}.$$

By Lagrange's Theorem, the only possible orders for non-trivial subgroups are 2, 3 and 6.

Any subgroup of order 2 consists of the identity and an element of order 2. Hence, the subgroups of order 2 are

$$\{e, (12)\}, \quad \{e, (13)\} \quad \text{and} \quad \{e, (23)\}.$$

None of these is normal. For example, the first is not normal, since taking the element (12) and conjugating by (13) gives

$$\begin{aligned} (13)(12)(13)^{-1} &= (13)(12)(13) \\ &= (13)(132) \\ &= (23) \notin \{e, (12)\}. \end{aligned}$$

Similar calculations show that neither of the other two is normal.

Any subgroup of order 3 is normal, because it has index 2. Any such subgroup must contain an element of order 3, and its inverse (also of order 3). Hence, the only subgroup of order 3 is

See Exercise 3.4 of *Unit IB4*.

$$\{e, (123), (132)\},$$

and as we saw above this subgroup must be normal.

The other normal subgroup is S_3 itself.

See Exercise 3.2 of *Unit IB4*.

(b) The possible quotient groups have orders $2 = 6/3$ and $1 = 6/6$. The quotient group of order 2 is cyclic. It is, therefore, Abelian. The trivial group, of order 1, is Abelian. Thus all quotients by non-trivial normal subgroups are Abelian, but S_3 is not.

Solution 2.7

Let $x, y \in G$. From the given information,
$$x^2 = y^2 = e.$$
Now, we also know that $(xy)^2 = e$, i.e.
$$\begin{aligned}
xyxy &= e \\
\Rightarrow x(xyxy) &= xe = x \\
\Rightarrow x^2 yxy &= x \\
\Rightarrow yxy &= x \\
\Rightarrow y(yxy) &= yx \\
\Rightarrow y^2 xy &= yx \\
\Rightarrow xy &= yx.
\end{aligned}$$

Solution 2.8

If $ba = e$ then $b = a^{-1} = a^2$, contradicting the fact that b is not in H.

If $ba = a$ then right cancellation gives the contradiction $b = e$.

If $ba = a^2$ then right cancellation gives the contradiction $b = a$.

Finally, if $ba = b$ then left cancellation gives the contradiction $a = e$.

Solution 2.9

We proceed by eliminating the other possible orders, namely 1 and 3.   *Remember that we are assuming that G has no element of order 6.*

Since b is not the identity element it does not have order 1.

Suppose that b has order 3, and hence that $b^4 = b$.
By cancellation, b^2 cannot be any of b, ab and $a^2 b$.
So b^2 must be one of e, a or a^2.
The element b^2 cannot be e because we have assumed that b has order 3.
Next, if $b^2 = a$, then
$$b = b^4 = a^2,$$
a contradiction (since b is not in H).
Finally, if $b^2 = a^2$, then
$$b = b^4 = a^4 = a.$$
This contradiction completes the proof.

Solution 2.10

We know that a has order 3 and that b has order 2. We also know that, since $ab = ba$,
$$\begin{aligned}
(ab)^6 &= a^6 b^6 \\
&= e^2 e^3 \\
&= e.
\end{aligned}$$

On the other hand, the non-identity element ab cannot have order 2 or 3 because
$$\begin{aligned}
(ab)^2 &= a^2 b^2 \\
&= a^2 e \\
&= a^2 \neq e
\end{aligned}$$
and
$$\begin{aligned}
(ab)^3 &= a^3 b^3 \\
&= eb \\
&= b \neq e.
\end{aligned}$$
Hence ab has order 6.

We know from Exercise 2.8 that $ba \neq e$; therefore, since $ab = ba$, $ab \neq e$.

Solution 2.11

(a) In the quotient group, the square of any element is the identity, hence
$$(bA)(bA) = b^2 A = A.$$
Therefore
$$b^2 \in A.$$

(b) The hint that $(b^2)^2 = e$ shows that the order of b^2 is at most 2. In A, the elements a and a^3 have order 4. Thus, b^2 cannot be a or a^3. Since $b^2 \in A$, the result follows.

Solution 2.12

If $ab = b$ then by right cancellation $a = e$, a contradiction.

If $ab = ba^2$ then
$$\begin{aligned} a^2 b &= a(ab) \\ &= a(ba^2) \\ &= (ab)a^2 \\ &= ba^2 a^2 \\ &= ba^4 \\ &= b. \end{aligned}$$

But by right cancellation this gives the contradiction $a^2 = e$.

Solution 2.13

(a) We calculate the powers of a:
$$a = \begin{bmatrix} i & 0 \\ 0 & -i \end{bmatrix};$$
$$a^2 = \begin{bmatrix} i^2 & 0 \\ 0 & (-i)^2 \end{bmatrix}$$
$$= \begin{bmatrix} -1 & 0 \\ 0 & -1 \end{bmatrix};$$
$$a^3 = \begin{bmatrix} -i & 0 \\ 0 & i \end{bmatrix};$$
$$a^4 = \begin{bmatrix} 1 & 0 \\ 0 & 1 \end{bmatrix}$$
$$= e.$$

Next:
$$b = \begin{bmatrix} 0 & -1 \\ 1 & 0 \end{bmatrix};$$
$$b^2 = \begin{bmatrix} -1 & 0 \\ 0 & -1 \end{bmatrix}$$
$$= a^2.$$

Finally:
$$ab = \begin{bmatrix} 0 & -i \\ -i & 0 \end{bmatrix} = ba^3.$$

(b) We already have distinct elements
$$e, a, a^2, a^3, b, ba^3.$$

The remaining elements
$$ba = \begin{bmatrix} 0 & i \\ i & 0 \end{bmatrix}$$

and
$$ba^2 = \begin{bmatrix} 0 & 1 \\ -1 & 0 \end{bmatrix}$$

are also distinct. Thus we have eight distinct elements.

(c) We present the Cayley table in terms of a and b:

	e	a	a^2	a^3	b	ba	ba^2	ba^3
e	e	a	a^2	a^3	b	ba	ba^2	ba^3
a	a	a^2	a^3	e	ba^3	b	ba	ba^2
a^2	a^2	a^3	e	a	ba^2	ba^3	b	ba
a^3	a^3	e	a	a^2	ba	ba^2	ba^3	b
b	b	ba	ba^2	ba^3	a^2	a^3	e	a
ba	ba	ba^2	ba^3	b	a	a^2	a^3	e
ba^2	ba^2	ba^3	b	ba	e	a	a^2	a^3
ba^3	ba^3	b	ba	ba^2	a^3	e	a	a^2

Hence, the set is closed. Matrix multiplication is associative. We have an identity, e, and an inverse for each element, as the table shows. Thus we have a group of order 8.

(d) Since the Cayley table is not symmetric about the leading diagonal, this matrix group is non-Abelian. It is, therefore, not isomorphic to any of the Abelian groups C_8, $C_2 \times C_4$ or $C_2 \times C_2 \times C_2$.

Reading down the leading diagonal of the table shows that a^2 is the only element of order 2 in this group. Since D_4 contains four reflections and a half-turn, it has five elements of order 2 and so cannot be isomorphic to this matrix group.

Solution 4.1

We must show that every element of G/H is of the form
$$(aH)^k$$

for some $k \in \mathbb{Z}$.

Suppose that xH is any element of G/H.
Then, since G is generated by a, it follows that
$$x = a^k$$

for some integer k. But
$$(aH)^k = (a^k)H$$
$$= xH.$$

Hence aH does generate G/H.

Solution 4.2

(a) Since m is the least positive exponent, we have $k \geq m$ and we can use the Quotient–remainder Theorem to write
$$k = mq + r, \quad 0 \leq r < m.$$
Hence
$$r = k - mq.$$
So,
$$\begin{aligned} a^r &= a^{k-mq} \\ &= a^k a^{-mq} \\ &= a^k (a^m)^{-q}. \end{aligned}$$
Since a^k and a^m belong to H, so does a^r.

Now, m was chosen as the least *positive* exponent appearing in H, so we must have $r = 0$, and hence
$$m \mid k.$$
We have shown that every element a^k of H is a power
$$a^k = a^{mq} = (a^m)^q$$
of a^m, which completes the proof that
$$H = \langle a^m \rangle.$$

(b) Since $a^n = e$, which is an element of H, from the previous part we have that
$$m \mid n.$$

(c) Let $n = mq$. We first observe that
$$e = a^n = a^{mq} = (a^m)^q.$$
So q is a positive power of a^m which produces the identity. It remains to prove that q is the smallest such power of a^m.

Consider a positive power k of a^m which is the identity, that is
$$e = (a^m)^k = a^{mk}.$$
Since the order of a is n, this can only be true if
$$mk \geq n = mq.$$
Hence
$$k \geq q.$$
So, q is the least power of a^m which gives the identity. Hence the order of a^m, and therefore the order of H, is q.

Solution 4.3

We do this by showing that the element a^m has order q.

We first observe that
$$e = a^n = a^{mq} = (a^m)^q.$$
So q is a positive power of a^m which produces the identity. It remains to show it is the smallest such power.

Now consider a positive power k of a^m which is the identity, that is
$$e = (a^m)^k = a^{mk}.$$
Since the order of a is n, this can only be true if
$$mk \geq n = mq.$$

Hence
$$k \geq q.$$
So, q is the least power of a^m which gives the identity.
Hence the order of a^m is q.

Solution 4.4

(a) The divisors of 24 are
$$1, 2, 3, 4, 6, 8, 12 \text{ and } 24,$$
so $\mathbb{Z}_{24}$ will have a subgroup of each of these orders.

(b) (i) The subgroups are isomorphic to
$$\mathbb{Z}_1, \mathbb{Z}_2, \mathbb{Z}_3, \mathbb{Z}_4, \mathbb{Z}_6, \mathbb{Z}_8, \mathbb{Z}_{12}, \mathbb{Z}_{24},$$
respectively.

(ii) The first subgroup in the list is the trivial subgroup generated by
$$24/1 \equiv 0 \pmod{24}.$$
The remaining ones have generators
$$12, 8, 6, 4, 3, 2, 1,$$
respectively.

We list the generators, the elements generated and the cyclic group to which the subgroup is isomorphic:

$24/1 \equiv 0$	$\{0\}$	$\mathbb{Z}_1$
$24/2 \equiv 12$	$\{0, 12\}$	$\mathbb{Z}_2$
$24/3 \equiv 8$	$\{0, 8, 16\}$	$\mathbb{Z}_3$
$24/4 \equiv 6$	$\{0, 6, 12, 18\}$	$\mathbb{Z}_4$
$24/6 \equiv 4$	$\{0, 4, 8, 12, 16, 20\}$	$\mathbb{Z}_6$
$24/8 \equiv 3$	$\{0, 3, 6, 9, 12, 15, 18, 21\}$	$\mathbb{Z}_8$
$24/12 \equiv 2$	$\{0, 2, 4, 6, 8, 10, 12, 14, 16, 18, 20, 22\}$	$\mathbb{Z}_{12}$
$24/24 \equiv 1$	$\{0, 1, \ldots, 23\}$	$\mathbb{Z}_{24}$

Solution 5.1

Following the argument in the text, the order of the element $(1,1)$ in the direct product will be the LCM of the orders of element 1 in each of the individual groups.

(a) The individual orders are 3 and 5.
Since $\operatorname{lcm}\{3,5\} = 15$, the element $(1,1)$ has order 15.
As $\mathbb{Z}_3 \times \mathbb{Z}_5$ has 15 elements, it is cyclic.

(b) The individual orders are 4 and 5.
Since $\operatorname{lcm}\{4,5\} = 20$, the element $(1,1)$ has order 20.
As $\mathbb{Z}_4 \times \mathbb{Z}_5$ has 20 elements, it is cyclic.

Solution 5.2

(a) The individual orders are 2 and 4.
Since $\operatorname{lcm}\{2,4\} = 4$ the element $(1,1)$ has order 4.
As $\mathbb{Z}_2 \times \mathbb{Z}_4$ has 8 elements, it is not generated by $(1,1)$.

All we have done is to show that $(1,1)$ is not a generator. To show that the direct product is not cyclic, we would have to show that *none* of the elements generates the whole group.

(b) If (a,b) is any element of the direct product, then a has order 1 or 2 and b has order 1, 2 or 4. The only possible LCMs of the orders of a and b are 1, 2 and 4. Hence no element of the direct product has order 8, and so $\mathbb{Z}_2 \times \mathbb{Z}_4$ is not cyclic.

Solution 5.3

In $\mathbb{Z}_6$, for any element a,
$$6a = 0.$$
In $\mathbb{Z}_8$, for any element b,
$$8b = 0.$$
Hence for any element (a, b) in $\mathbb{Z}_6 \times \mathbb{Z}_8$, we have
$$\begin{aligned}24(a,b) &= (24a, 24b) \\ &= (4(6a), 3(8b)) \\ &= (0,0).\end{aligned}$$
So, the maximum order for any element is at most 24. Therefore this group of order 48 cannot be cyclic.

We could reach exactly the same conclusion by considering the possible LCMs of the orders of a and b. Note that the maximum order, 24, is the LCM of 6 and 8.

Solution 5.4

(a) The group $\mathbb{Z}_4 \times \mathbb{Z}_6$ is not cyclic.

In $\mathbb{Z}_4$, for any element a,
$$4a = 0.$$
In $\mathbb{Z}_6$, for any element b,
$$6b = 0.$$
Hence for any element (a,b) in $\mathbb{Z}_4 \times \mathbb{Z}_6$, we have
$$\begin{aligned}12(a,b) &= (12a, 12b) \\ &= (3(4a), 2(6b)) \\ &= (0,0).\end{aligned}$$
So, the maximum order for any element is at most 12. Therefore the direct product, of order 24, is not cyclic.

(b) The group $\mathbb{Z}_2 \times \mathbb{Z}_9$ is cyclic.

Previous examples suggest that $(1,1)$ should be a generator, which we can confirm as follows.

The order of 1 in $\mathbb{Z}_2$ is 2 and the order of 1 in $\mathbb{Z}_9$ is 9.
Since $\mathrm{lcm}\{2,9\} = 18$, the element $(1,1)$ has order 18 in the direct product.
As the direct product has 18 elements, it is cyclic.

Solution 5.5

Suppose that
$$\mathrm{hcf}\{m,n\} = d > 1$$
and that
$$\mathrm{lcm}\{m,n\} = l.$$
Since
$$mn = dl$$
and $d > 1$, we have $l < mn$.

Now, l is a multiple of m and of n, so
$$l = mq_1 = nq_2,$$
say.

$d > 1$ since m and n are not coprime.

Hence, for any element (a, b) of the direct product, we have

$$l(a, b) = (la, lb)$$
$$= (q_1(ma), q_2(nb))$$
$$= (q_1 \times 0, q_2 \times 0)$$
$$= (0, 0).$$

This shows that the order of (a, b) cannot exceed l, which is *less* than mn, the order of the direct product.

Hence no element can generate the whole of the direct product, and so $\mathbb{Z}_m \times \mathbb{Z}_n$ is not cyclic.

Solution 5.6

We write

$$90 = 2 \times 45.$$

Since 2 and 45 are coprime, we can deduce from Lemma 5.1 that

$$\mathbb{Z}_{90} \cong \mathbb{Z}_2 \times \mathbb{Z}_{45}.$$

Now we can repeat this idea with $\mathbb{Z}_{45}$.
Since $45 = 9 \times 5$ and these factors are coprime, we have

$$\mathbb{Z}_{45} \cong \mathbb{Z}_9 \times \mathbb{Z}_5.$$

Combining these results, we have

$$\mathbb{Z}_{90} \cong \mathbb{Z}_2 \times \mathbb{Z}_{45}$$
$$\cong \mathbb{Z}_2 \times \mathbb{Z}_9 \times \mathbb{Z}_5.$$

Solution 5.7

We generalize the argument used in the last solution. We write

$$n = \left(p_1^{k_1}\right)\left(p_2^{k_2} \ldots p_r^{k_r}\right)$$
$$= n_1 \left(n_2 \ldots n_r\right),$$

where, because all the primes are distinct, the two factors are coprime.
Hence, by Lemma 5.1,

$$\mathbb{Z}_n \cong \mathbb{Z}_{n_1} \times \mathbb{Z}_{n_2 \ldots n_r}.$$

Now, repeating the argument another $r - 1$ times, at each stage separating one term, we obtain the required result.

Note that any proof that uses a phrase such as 'repeating the argument ...' has the Principle of Mathematical Induction hidden in it. In this case a formal proof by induction would use induction on the number of distinct prime factors, noting that for one prime factor the result is trivially true.

Solution 5.8

We apply the result of the last exercise to the various groups appearing in each direct product. For example, since

$$154 = 2 \times 7 \times 11,$$

we have

$$\mathbb{Z}_{154} \cong \mathbb{Z}_2 \times \mathbb{Z}_7 \times \mathbb{Z}_{11}.$$

Hence

$$\mathbb{Z}_{154} \times \mathbb{Z}_{20} \times \mathbb{Z}_5 \cong (\mathbb{Z}_2 \times \mathbb{Z}_7 \times \mathbb{Z}_{11}) \times (\mathbb{Z}_4 \times \mathbb{Z}_5) \times \mathbb{Z}_5$$
$$\cong \mathbb{Z}_2 \times \mathbb{Z}_4 \times \mathbb{Z}_5 \times \mathbb{Z}_5 \times \mathbb{Z}_7 \times \mathbb{Z}_{11}.$$

Similarly

$$\mathbb{Z}_{55} \times \mathbb{Z}_{28} \times \mathbb{Z}_{10} \cong (\mathbb{Z}_5 \times \mathbb{Z}_{11}) \times (\mathbb{Z}_4 \times \mathbb{Z}_7) \times (\mathbb{Z}_2 \times \mathbb{Z}_5)$$
$$\cong \mathbb{Z}_2 \times \mathbb{Z}_4 \times \mathbb{Z}_5 \times \mathbb{Z}_5 \times \mathbb{Z}_7 \times \mathbb{Z}_{11}.$$

Thus the two groups are isomorphic to the same direct product and hence to one another.

OBJECTIVES

After you have studied this unit, you should be able to:

(a) apply the type of argument used to classify groups of order up to 8;

(b) given a finite cyclic group, find a generator for a subgroup of given order;

(c) express a given finite cyclic group as the direct product of cyclic groups of prime power order;

(d) given two direct products of cyclic groups, determine whether or not they are isomorphic;

(e) prove simple results about Abelian groups, cyclic groups and direct products.

INDEX

Abelian group 13
classification of cyclic groups 25
decomposition of $\mathbb{Z}_{mn}$ 31
decomposition theorem
    for finite cyclic groups 31
direct products of cyclic groups 30
Hamilton's quaternion group 19
internal direct product 9
internal direct product theorem 9
non-trivial subgroup 14
proper subgroup 14
quaternion group 19
quotients of cyclic groups 25
subgroups of $\mathbb{Z}$ 25
subgroups of cyclic groups 26
$\mathbb{Z}_n$ 24